How Does It Grow?

HOW DO CARROTS GROW?

By Kathleen Connors

Gareth Stevens
PUBLISHING

Please visit our website, www.garethstevens.com. For a free color catalog of all our high-quality books, call toll free 1-800-542-2595 or fax 1-877-542-2596.

Library of Congress Cataloging-in-Publication Data
Names: Connors, Kathleen, author.
Title: How do carrots grow? / Kathleen Connors.
Description: New York : Gareth Stevens Publishing, [2022] | Series: How does it grow? | Includes index.
Identifiers: LCCN 2020011771 | ISBN 9781538262993 (library binding) | ISBN 9781538262979 (paperback) | ISBN 9781538262986 (6 Pack) | ISBN 9781538263006 (ebook)
Subjects: LCSH: Carrots–Juvenile literature.
Classification: LCC SB351.C3 C67 2022 | DDC 635/.13–dc23
LC record available at https://lccn.loc.gov/2020011771

First Edition

Published in 2022 by
Gareth Stevens Publishing
111 East 14th Street, Suite 349
New York, NY 10003

Editor: Kristen Nelson
Designer: Katelyn E. Reynolds

Photo credits: Cover, p. 1 Bill Sykes/Cultura/Getty Images; p. 5 Kevin Summers/ Photographer's Choice / Getty Images Plus; pp. 7, 24 (seeds) Heike Rau/Shutterstock.com; pp. 9, 24 (soil) Geography Photos/ Universal Images Group via Getty Images; p. 11 Stuart Fox/ Gallo Images / Getty Images Plus; p. 13 PhotoAlto/Laurence Mouton/ PhotoAlto Agency RF Collections/Getty Images; p. 15 Yuji Sakai/ DigitalVision/ Getty Images; p. 17 Victoria ArtWK/ iStock / Getty Images Plus; p. 19 emholk/E+/Getty Images; p. 21 sergio_kumer/ iStock / Getty Images Plus; p. 23 Mallivan/ iStock / Getty Images Plus.

Printed in the United States of America

CPSIA compliance information: Batch #CSGS22: For further information contact Gareth Stevens, New York, New York at 1-800-542-2595.

Contents

Carrots taste good.
How do they grow?

They grow from seeds.
The seeds are tiny!

Seeds are planted in soil.
They grow well in rows.

They need water.
They need sunlight.

Leaves come out of the soil.

The carrot is the root.
It grows underground.

Carrots start growing in spring.

We dig them up in fall.
This is the harvest.

They may be big or small.

They can be orange.
They can be yellow!

Words to Know

seeds

soil

Index